Parasitology

For CLS & MLT

Mary Michelle Shodja, MS, MT (ASCP), CLS

Order this book online at www.trafford.com
or email orders@trafford.com

Most Trafford titles are also available at major online book retailers.

Printed in the United States of America.

ISBN: 978-1-4269-6173-1

Trafford rev. 07/19/2011

Trafford
PUBLISHING®   www.trafford.com

North America & international
toll-free: 1 888 232 4444 (USA & Canada)
phone: 250 383 6864 ♦ fax: 812 355 4082

Contents

List of Figures iii

List of Pictures. iv

List of Tables vi

Preface viii

SECTION I 1

Introduction to Parasitology

SECTION II. 9

Sources & Common Opportunistic

Pathogens Isolated

SECTION III 13

Parasitology Specimen Processing

SECTION IV 21

Parasite Identification

SECTION V 41

Medically Important Parasites

SECTION VI Treatment 45

Parasitology Written Examination. 48

Parasitology Written Exam Answer Key 55

Afterword 56

Glossary 57

References 60

About the Author 61

List of Figures

Figure 1 – 2 Major Taxonomic Group of Parasites 1

Figure 2 – Medically Important Protozoans 2

Figure 3 – Medically Important Helminths 5

List of Pictures

Picture 1 - Diagram on how to scan parasites in a microscopic field. 21

Picture 2 – Plasmodium falciparum multiple rings 24

Picture 3 - Plasmodium falciparum gametocyte 24

Picture 4 – Plasmodium vivax 25

Picture 5 - Plasmodium ovale. 25

Picture 6 - Plasmodium malariae. 26

Picture 7 -Babesia microti 27

Picture 8 – L. donovani amastigote (in tissue) 28

Picture 9 – L. donovani promastigote (in culture) 28

Picture 10 – L. donovani (in blood inside WBC). 29

Picture 11 – Toxoplasma gondii in WBC (arrow) 29

Picture 12 - T. gondii tachyzoites released from tissues 30

Picture 13 - Trypanosoma cruzi trypomastigote 30

Picture 14 - Trypanosoma brucei gambiense 31

Picture 15 – Wucherera bancrofti 31

Picture 16 – Loa loa 32

Picture 17 – Entamoeba hystolitica (cyst) 32

Picture 18 – Entamoeba coli (cyst) 32

Picture 19 – Endolimax nana (cyst) 33

Picture 20 - Iodamoeba butschlii (cysts) 33

Picture 21 - Blastocystis hominis 33

Picture 22 - Giardia lamblia (cyst) 33

Picture 23 - Giardia lamblia (troph) trichrome 33

Picture 24 – Chilomastix mesnili (cyst) 34

Picture 25 - ientamoeba fragilis   troph –trichrome 34

Picture 26 - Balantidium coli(troph) 34

Picture 27 – Isospora belli (oocyst) 34

Picture 28 – Ascaris lumbricoides (fertilized corticoid) 34

Picture 29 - Ascaris lumbricoides (fertilized decorticoid) 35

Picture 30 - Ascaris lumbricoides (unfertilized corticoid) 35

 Picture 31 - Hookworm eggs 35

Picture 32 - Enterobius vermicularis Pinworm 35

Picture 33 - Strongyloides stercolaris (larva) 35

Picture 34 - Trichuris trichiura -Whip worm 35

Picture 35 – Tapeworm egg 36

Picture 36 - Diphyllobotrium latum (Fish tapeworm) 36

Picture 37 - Hymenolepis nana – Dwarf tapeworm 36

Picture 38 - Schistosoma haematobium 37

Picture 39 - Schistosoma japonicum 37

Picture 40 - Schistosoma mansoni 37

Picture 41 - Paragonimus westermani 37

Picture 42 - Clonorchis sinensis 37

Picture 43 - Fasciolopsis buski 37

Picture 44 - Fasciola hepatica 38

Picture 45 - Echinococcus granulosus 38

List of Tables

Table 1 – Sources and Common Opportunistic Pathogens isolated 13

To my mother, for her sacrifices, this is for you.

To To my oldest sister Mae Solis, one of the hardest working individual I know, a woman of strength and independence, thank you for your love and support through the years. With deep gratitude to my cousins Chona Aros and Christine Sy. Lastly, to my classmate and best friend, Anna Hamilton, thank you for your friendship of over 20 years.

"Many of the things you can count, don't count. Many of the things you can't count, really count"
Albert Einstein

To laboratory students and fellow laboratorians, never forget what you learned and never stop learning new things
The Author

Preface

This manual is the third of a series of 20 manuals that the author was commissioned to write for the Medical Laboratory Technician (MLT) training program in Diamond Bar, California that was granted approval as a training facility in 2009 by the State Department of Health and Human Services. In writing these manuals, the author strived to adhere to the strict guidelines of the State of California's requirements for the MLT program. At the time of publication, the initial beneficiary of these manuals, Diamond Bar California's first MLT graduate, successfully passed the ASCP examination.

The author revised these manuals to serve 3 purposes: as the primary textbooks for the 6-month MLT training program, as reviewers for Clinical Laboratory Scientists (CLSs) preparing for the CLS Licensure or Certification, and as continuing education materials for CLSs and MLTs as a requirement for license renewal.

The author wrote these manuals in an outline form for easy reading and understanding and free from the constraint of a formal textbook. The author's intention is to speak to the reader from the actual clinical laboratory bench than from the classroom.

The Parasitology for CLS and MLT only discussed the most common parasites isolated in the clinical laboratory and is not recommended as a replacement of the actual textbooks currently being used in the CLS program.

Also by Mary Michelle Shodja, MS, MT (ASCP), CLS:

Bacteriology for CLS & MLT
Hematology for CLS & MLT
Mycology for CLS & MLT
Virology for CLS & MLT
Coagulation for CLS & MLT
Urinalysis & Body Fluids for CLS & MLT
Routine Chemistry for CLS & MLT
Special Chemistry for CLS & MLT
Toxicology for CLS & MLT
Serology - Immunoassays for CLS & MLT
Serology - ELISA Assays for CLS & MLT
Serology & Syphilis for CLS & MLT
Immunology for CLS & MLT
Blood Bank for CLS & MLT
Phlebotomy for CLS, MLT & Phlebotomist
Specimen Processing for CLS, MLT & Phlebotomist
Laboratory Safety for CLS, MLT & Phlebotomist
Total Quality Management I – Quality Assurance
Total Quality Management II – Quality Control

SECTION I
Introduction to Parasitology

Parasitology

- Study of parasites, their hosts and the relationship between them
- Medical Parasitology is the study of those parasites that infect humans
- Parasite – refers to organisms belonging to one or two major taxonomic groups:

Figure 1 – 2 Major Taxonomic Group of Parasites

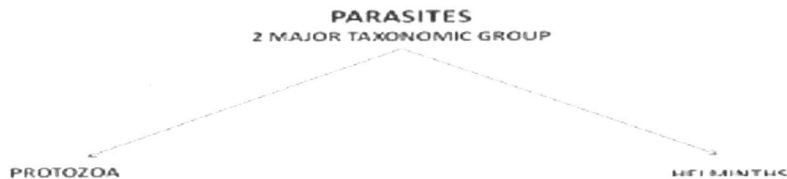

PARASITES
2 MAJOR TAXONOMIC GROUP

PROTOZOA   HELMINTHS

a. Protozoa – microscopic, single-celled eukaryocytes superficially resembling yeasts in both size and simplicity
b. Helminths – macroscopic, multicellular worms possessing differentiated tissues and complex organ systems varying in length from a meter to a less than a millimeter

CHARACTERISTICS OF PROTOZOA AND HELMINTHS
- Majority are free-living
- Majority play a significant role in the ecology of the planet
- Majority seldom inconvenience the human race
- The less common disease-producing species are typically obligate parasites, dependent on vertebrate hosts, arthropod hosts or both for survival
- Parasite level of adaptation with the host is high = presence in the host produce little or no injury
- Parasite level of adaptation with the host is low (less complete adaptation) = more serious disturbance of the host
- Encystment – cyst – protective capsule, permits organism to survive when food, oxygen or moisture is lacking, allows parasite to survive outside of the host

1

Figure 2 – Medically Important Protozoans

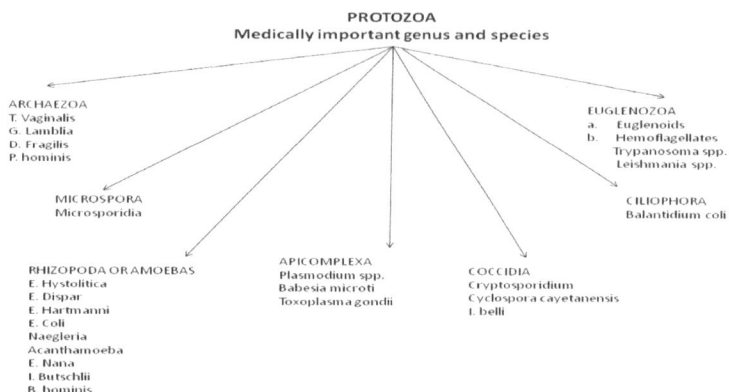

PROTOZOA
Medically important genus and species

ARCHAEZOA
T. Vaginalis
G. Lamblia
D. Fragilis
P. hominis

EUGLENOZOA
a.   Euglenoids
b.   Hemoflagellates
     Trypanosoma spp.
     Leishmania spp.

MICROSPORA
Microsporidia

CILIOPHORA
Balantidium coli

RHIZOPODA OR AMOEBAS
E. Hystolitica
E. Dispar
E. Hartmanni
E. Coli
Naegleria
Acanthamoeba
E. Nana
I. Butschlii
B. hominis

APICOMPLEXA
Plasmodium spp.
Babesia microti
Toxoplasma gondii

COCCIDIA
Cryptosporidium
Cyclospora cayetanensis
I. belli

PROTOZOA
Medically important Phyla of Protozoa

1.   Archaezoa – Flagellates - lack mitochondria, many live as symbionts in the digestive tract of animals, spindle-shaped with flagella (2 or more) from front end

   o   Trichomonas vaginalis –no cyst stage so must be transferred from host-to-host, found in the vagina and male urinary tract. Transmitted sexually or through toilet facilities or towels
   o   Giardia lamblia
   o   Dientamoeba fragilis
   o   Penatrichomonas (Trichomonas) hominis

2.   Microspora – lack mitochondria, no microtubules, obligate intracellular parasites, cause of chronic diarrhea, keratoconjunctivitis
   o   Microsporidia

3.   Rhizopoda or amoebas – move by extending blunt, lobelike projections of the cytoplasm (pseudopods)
   o   Entamoeba hystolitica – only pathogenic amoeba in human intestine, identified through DNA analyses and lectin binding, causes amoebic dysentery, transmitted between humans through ingestion of cyst

NON-PATHOGENIC SPECIES IN HUMAN INTESTINE
- o Entamoeba dispar
- o Entamoeba hartmanni
- o Entamoeba coli
- o Endolimax nana
- o Iodamoeba butschlii
- o Blastocystis hominis
  OTHER PATHOGENIC AMOEBA
- o Naegleria (Primary Amoebic Menigoencephalitis-PAM)
- o Acanthamoeba – grows in water, causes corneal infection and causes blindness

4.   Apicomplexa (sporozoa) – not motile in mature forms, obligate intracellular, characterized by presence of complex organelles at the apexes (tips) of their cells containing enzymes that can penetrate host cells, has complex life cycle that involves transmission between hosts
- o Plasmodium – causes malaria
  - ➢ Mosquito (definitive host) –Anopheles mosquito
  - ➢ Humans (intermediate host)

- o Babesia microti – causes fever and anemia
  - ➢ transmitted by Ixodes scapularis (tick), life cycle involves domestic cats

- o Toxoplasma gondii – intracellular parasite
  - ➢ tachyzoites (reproduce sexually and asexually) in infected cats
  - ➢ ingested by humans as tachyzoites

5.   Coccidia
- o Cryptosporidium – infects immunocompromised transmitted to humans through feces of cows, rodents, dogs, cats
  - ➢ Cryptosporidium parvum
- o Cyclospora cayetanensis – causes water-borne diarrhea
- o Isospora belli

6.   Ciliophora – has cilia (similar to but shorter than flagella)
   - o   Balantidium coli – only ciliate that is a human parasite, causes severe but rare type of dysentery, ingestion of cyst

7.   Euglenozoa – 2 groups of flagellated cells based on rRNA sequences, disk-shaped mitochondria and absence of sexual production

   a.   Euglenoids – photoautotrophs, has pellicles, flagella at anterior end, most have red eyespots at anterior end that sense light and directs cell in appropriate direction by pre-emergent flagellum
   b.   Hemoflagellates – (blood parasites) – transmitted by bites of blood-feeding insects, found in circulatory system of bitten host, have long, slender bodies with undulating membrane

      - o   Trypanosoma
        - ➤ T. brucei gambiense - causes African sleeping sickness transmitted by tsetse fly
        - ➤ T. cruzi - agent of Chagas disease, transmitted by "kissing bug" bites on the face and defacates on the wound releasing trypanosomes
      - o   Leishmania
        - ➤ Leishmania donovani – causes disseminated visceral (kala azar) disease
        - ➤ Leishmania braziliensis – causes mucocutaneous lesions (espundia)

Figure 3 – Medically Important Helminths

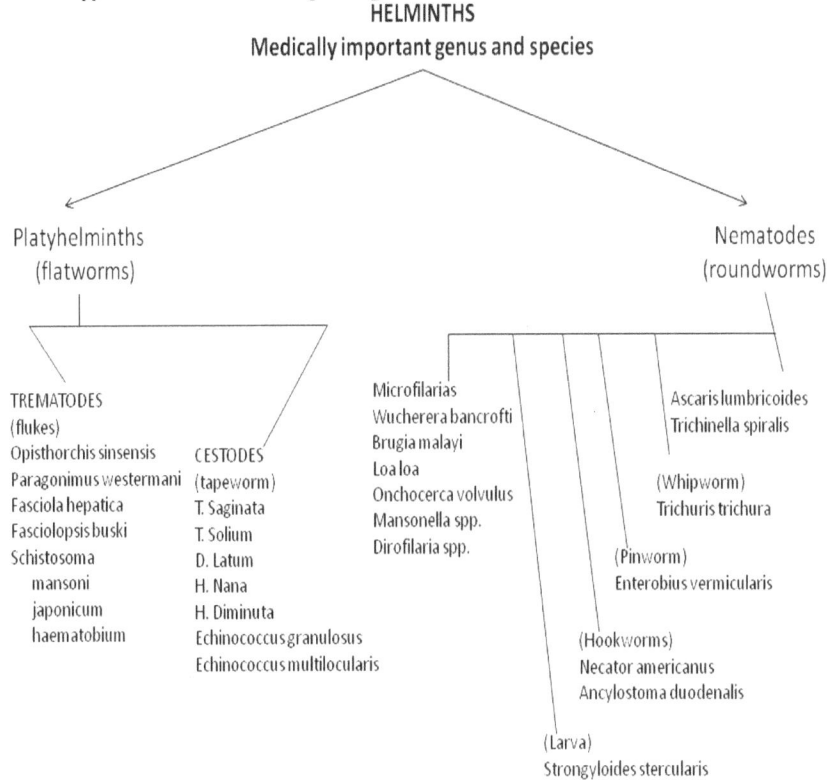

HELMINTHS
Medically important genus and species

```
                          HELMINTHS
               Medically important genus and species
                    /                        \
    Platyhelminths                              Nematodes
    (flatworms)                                 (roundworms)
```

TREMATODES
(flukes)
Opisthorchis sinsensis
Paragonimus westermani
Fasciola hepatica
Fasciolopsis buski
Schistosoma
   mansoni
   japonicum
   haematobium

CESTODES
(tapeworm)
T. Saginata
T. Solium
D. Latum
H. Nana
H. Diminuta
Echinococcus granulosus
Echinococcus multilocularis

Microfilarias
Wucherera bancrofti
Brugia malayi
Loa loa
Onchocerca volvulus
Mansonella spp.
Dirofilaria spp.

Ascaris lumbricoides
Trichinella spiralis

(Whipworm)
Trichuris trichura

(Pinworm)
Enterobius vermicularis

(Hookworms)
Necator americanus
Ancylostoma duodenalis

(Larva)
Strongyloides stercularis

HELMINTHS

- Larval – developmental stage
- Adult may be dioecious – male and female, OR monoecious (hermaphroditic) reproduction occurs when both are present in the same host

Medically Important Phyla of Helminths

1. Platyhelminths (flatworms)
   o Trematodes (flukes) – has ventral and oral suckers, absorbs food through cuticle (lung fluke, liver fluke, blood fluke)
     ➢ Opisthorchis (Clonorchis) sinensis – Asian (Chinese) liver fluke
     ➢ Paragonimus westermani- lung fluke
     ➢ Fasciolopsis buski – giant intestinal fluke

- ➢ Fasciola hepatica – sheep liver fluke
- ➢ Schistosoma – blood fluke
  - a. Schistosoma mansoni – infects veins over large instestine
  - b. Schistosoma haematobium – infects veins over bladder
  - c. Schistosoma japonicum – infectis veins over small intestine

- o Cestodes (tapeworms) – intestinal and tissue parasites, has scolex (head) which has suckers and some have small hooks, absorbs food through cuticles, has body that consists of segments (proglottids). Humans are definitive host.
  - ➢ Taenia saginata - beef tapeworm
  - ➢ Taenia solium - pork tapeworm
  - ➢ Diphyllobotrium latum - freshwater fish tapeworm
  - ➢ Hymenolepis nana – dwarf tapeworm
  - ➢ Hymenolepis diminuta – rat tapeworm
  - ➢ Echinococcus granulosus – unilocular cyst
  - ➢ Echinococcus multilocularis – metastatic spreading cysts

2.   Nematoda (round worms) cylindrical and tapered at ends, has complete digestive system (mouth, intestine and anus), most are dioecious, male smaller than female, have 1 or 2 spicules on posterior end, infections are either through eggs or larva
- ➢ Enterobious vermicularis – pin worm
- ➢ Ascaris lumbricoides – large nematode, dioecious, most common infection worldwide
- ➢ Necator americanus and Ancylostoma duodenalis – hookworms
- ➢ Trichinella spiralis – causes trichinosis
- ➢ Trichuris trichiura – whipworm
- ➢ Strongyloides stercolaris – hyperinfection causes visceral, ocular and larva migrans
- ➢ Anisakines or wriggly worms – infects fish
- ➢ Filarial worms:
  Wuchereria bancrofti

  Brugia malayi

  Brugia timori

  Loa loa

  Onchocerca volvulus

  Mansonella ozzardi

  Mansonella streptocerca

Mansonella perstans
Dirofilaria immitis – lung lesion in dogs
Dirofilaria spp. – may be found in subcutaneous nodules

ARTHROPODS AS VECTORS

• Arthropods – animals characterized by segmented bodies, hard external skeletons and jointed legs, >1 million species, largest phylum in the animal kingdom, few suck the blood of humans and animals and can transmit microbial disease

• Vectors – arthropods that can carry pathogenic microorganisms
1. Arachnia (8 legs) – spiders, mites, ticks, scorpions
2. Crustacea (4 antennae) – crabs, crayfish, copepods
3. Pentastomida – tongue worms
4. Dipolopoda – millipedes
5. Chilopoda – centipedes
6. Insecta – bees, flies, lice
   o Insects and ticks reside on the animal only when feeding except louse (spends entire time on the host)
   o Some vectors are mechanical only
   o Some parasites multiply on vectors (accumulate on vector's feces) – Lyme disease and West Nile
   o Vectors and definitive host – Plasmodium

SECTION II
Sources & Common Opportunistic Pathogens Isolated

Table 1 - Sources and Common Pathogens Isolated

Source	Common Pathogens Isolated
Intestinal	Entamoeba hystolytica, coli and hartmanni Iodamoeba butschlii Blastocystis hominis Giardia lamblia Chilomastix mesnili Dientamoeba fragilis Pentatrichomonas hominis Balantidium coli Cryptosporidium parvum Isopora belli Microscporidia Ascaris lumbricoides Enterobius vermicularis Hookworms Strongyloides stercularis Trichuris trichura Hymenolepis nana, diminuta Taenia saginata, solium Dyphyllobotrium latum Clonorchis sinensis Paragonimus westermani Schistosoma spp. Fasciolopsis buski Fasciola hepatica Metagnonimus yokagawai Heterophyes heterophyes

Muscle	Taenia solium, Trichinella spiralis, Onchocerca volvulus, Trypanosoma cruzi, Microsporidia

Blood	Red Blood Cells – Plasmodium spp. & Babesia spp. White Blood Cells – Leishmania donovani & Toxoplasma gondii Whole Blood – Trypanosoma spp. & Microfilariae Bone Marrow – Leishmania donovani
Eye	Acanthamoeba spp., Toxoplasma gondii, Loa loa, Microsporidia
Liver/Spleen	Echinococcus spp., Entamoeba hystolitica, Leishmania donovani, Microsporidia
Lung	Cryptosporidium spp., Echinococcus spp., Paragonimus spp., Microsporidia
Central Nervous System	Taenia solium, Echinococcus, Naegleria fowleri, Acanthamoeba/Hartmanella spp., Toxoplasma gondii, Microsporidia, Trypanosoma spp.
Cutaneous ulcer	Leishmania spp.
Skin	Leishmania spp., Onchocerca volvulus, Microfilariae
Urogenital	Trichomonas vaginalis, Schistosoma spp., Microsporidia

SECTION III
Parasitology Specimen Processing

SPECIMEN PRESERVATION
- Preservatives – will prevent the deterioration of any parasites that are present, useful when fecal specimen cannot be processed and examined in the recommended time
- Fecal
    - Polyvinyl Alcohol (PVA) – plastic resin that is normally incorporated into Schaudinn's fixative (mercuric chloride with distilled water). PVA powder serves as an adhesive for the stool material so when it is spread onto the slide, it sticks to it. Highly recommended for preserving cysts and trophozoites. Ratio of fecal to preservative should be 1:3.
    - Modified PVA – copper-based (no mercury) - does not provide good preservation of protozoan morphology
    - Modified PVA – zinc-based (no mercury) – still not as beautifully preserved as mercury-based but organisms could be clearly identified
    - Sodium Acetate-Acetic Acid Formalin (SAF) – lends itself to both the concentration technique and permanent stained smear and does not contain mercury. Liquid fixative much like 10% formalin and considered a "softer" fixative than mercuric chloride. Morphology of organisms on permanent stain is not as sharp as the mercuric chloride preserved specimen but staining SAF with iron-hematoxylin is better than staining with trichrome stain
    - Merthiolate (thimerosal)-iodine-formalin (MIF) – good preservatives for most kinds and stages of parasites
    - 5-10% Formalin – all-purpose fixative that is appropriate for helminth eggs and larvae and protozoan cysts
        - 5% - recommended for preservation of protozoan cysts
        - 10% - recommended for helminth eggs and larvae, commonly used
- Blood – EDTA

SPECIMEN PROCESSING
1.   Direct Wetmount – to assess the worm burden of a patient, to provide a quick diagnosis of heavily infected specimen, to check organism motility, and to diagnose parasites that may be lost in concentration techniques

- - Saline 0.85% - useful in scanning for cysts and eggs and motility of the trophozoites
  - Iodine (D'Antoni's or Lugol's) – useful for identification of cysts and eggs but the trophozoites are killed (no motility observed). The chromatin material of amoebic cysts stain dark brown against the yellow-brown cytoplasm, the nuclear structures are differentiated and the glycogen masses stain mahogany brown.
  - D'Antoni's Iodine – 1.0 g Potassium Iodide (KI), 1.5 g powdered iodine crystals, 100 ml distilled water
  - Lugol's Iodine – 10.0 g KI, 5.0 g powdered iodine crystals, 100 ml distilled water

- Procedure:
  a. Place 1 drop of 0.85% NaCl on the left side of the slide and 1 drop of iodine on the right side of the slide
  b. Take a very small amount of fecal specimen (about the amount picked up on the end of an applicator stick) and thoroughly emulsify the stool in the saline and iodine preparations (use separate stick for each)
  c. Place a coverslip (22x22 mm) on each suspension
  d. Scan the entire coverslip for both suspension under 10x magnification
  e. At least 1/3 of the coverslip should be examined with the 40x objective. Any suspicious structures should be verified under the 40x magnification

2. Concentrations (Sedimentation or flotation) – design to separate protozoan organisms and helminth eggs and larvae from fecal debris.
   - Formalin-ethyl acetate sedimentation
     - concentration by centrifugation procedure leads to the recovery of all protozoa, eggs and larvae present but there is more debris than the flotation procedure
     - Ethyl acetate is used as an extractor of debris and fat from the feces and leaves the parasites at the bottom of the suspension
     - Formalin-ethyl acetate sedimentation is the easiest to perform, allows recovery of the broadest range of organisms, and is least subject to technical error
     - Recommended for modified acid fast stain and modified trichrome stain used for coccidian and microsporidia
     - Procedure:
       a. Transfer a half-teaspoon (approximately 4 g) of fresh stool into a 10 ml of 10% formalin in a conical tube

b. Mix the stool and formalin thoroughly and let stand for 30 minutes for fixation
c. If the stool was already submitted in formalin skip a and b
d. Strain approximately 8 ml of the stool-formalin mixture through a wet gauze and into a 15ml conical tube
e. Add 0.85% NaCl almost to the top of tube and centrifuge for 10 minutes at 500 x g. Amount of sediment obtain should be 0.5-1.0 ml. Repeat if the supernatant fluid remains darker than tan
f. Decant the supernatant fluid and resuspend the sediment on the bottom of the tube in 10% formalin until half of the tube
g. Add 4-5 ml of ethyl acetate. Stopper the tube and shake it vigorously for at least 30 seconds, exerting pressure on the stopper throughout
h. Wait for 15-30 seconds and carefully remove the stopper
i. Centrifuge for 10 minutes at 500 x g. 4 layers should result: bottom-small amount of sediment, 2nd from the bottom layer-formalin, 3rd from the bottom layer-debris and fat, and the last layer – ether substitute
j. Decant all the supernatant fluid
k. Mix the sediment (add 1-2 drops of NaCl if sediment is still solid)
l. Examine under wetmount or perform a modified acid fast or trichrome stain

- o Zinc sulfate flotation – permits the separation of protozoan cysts and certain helminth eggs from excess debris through the use of a liquid with a high specific gravity

  - ➢ The parasitic elements are recovered in the surface film and debris remains in the bottom of the tube
  - ➢ This technique yields a cleaner preparation than the sedimentation procedure but some helminth eggs, operculated eggs and/or very dense eggs such as unfertilized Ascaris eggs) do not concentrate well with the flotation method
  - ➢ This technique is not commonly used in most laboratories
  - ➢ Procedure:
  - a. Transfer a half-teaspoon (approximately 4 g) of fresh stool into a 10 ml of 10% formalin in a conical tube
  - b. Mix the stool and formalin thoroughly and let stand for 30 minutes for fixation
  - c. If the stool was already submitted in formalin skip a and b
  - d. Add 0.85% NaCl almost to the top of the tube and centrifuge for 10 minutes at 500 x g

e. Perform a second wash by decanting the supernatant and repeating step d
f. Decant the supernatant and resuspend the sediment with 1-2 ml of zinc sulfate, then add zinc sulfate to within 2-3 mm of the rim
g. Centrifuge for 1 minute at 500 x g
h. 2 layers should be observed: bottom layer will have the debris and some operculated/heavy eggs, on the top film – contains the protozoan cysts and some helminth eggs
i. Without removing the tube from the centrifuge, take 1-2 drops of the surface film using a sterile disposable transfer pipette and place it on a slide
j. Add a coverslip or iodine can be added
k. Examine the specimen for ova and parasite under 10x magnification
l. All suspicious structures should be observed at 40x magnification

3. Permanent Slide
   ➢ single most productive means of stool examination for intestinal protozoa
   ➢ permanent stained smear facilitates detection and identification of cysts and trophozoites and affords a permanent record of the protozoa encountered
   ➢ small protozoa missed by direct smear and concentration techniques are often seen on the stained smear
   ➢ permanent stains also allows laboratories to refer the slide to a specialist for consultation of organisms with unusual morphology

   o Trichrome – technique of Wheatley for fecal specimens is a modification of Gomori's original staining procedure for tissue. It is a rapid, simple procedure that produces uniformly well-stained smears of intestinal protozoa, human cells, yeast cells, and artifact material

      ➢ Procedure:
         a. Fresh specimen– take a small fecal sample and make a smear on 2 slides and immediately place the slides on the Schaudinn's fixative for 30 minutes. Proceed to trichrome staining (step c)
         b. PVA/Modified PVA samples – mix the sample and pour onto a paper towel and let stand for 3 minutes (to absorb the PVA), take some of the sample and smear it onto 2 slides and allow to dry for several hours. Place the slides into iodine-alcohol for 1 minute (for PVA specimens to remove the mercury).

Proceed to trichrome staining (step c)

   c. Trichrome staining: Place the slides in 70% ethanol + iodine for 5 minutes

   d. Place slides in 70% ethanol for 5 minutes

   e. Place the slides in 70% ethanol again for 3 minutes

   f. Place in the trichrome stain for 10 minutes

   g. Place in 90% ethanol plus acetic acid for 1-3 seconds

   h. Dip several times in 100% ethanol for 3 minutes each

   i. Place in xylene for 5-10 minutes

   j. Mount with coverslip using a mounting medium

   k. Allow the smear to dry and examine for protozoan trophs and cysts under oil 100x magnification

- Iron Hematoxylin – stain used for most of the original morphological descriptions of intestinal protozoa found in humans

  - Procedure:
    a. Fresh specimen– take a small fecal sample and make a smear on 2 slides and immediately place the slides on the Schaudinn's fixative for 30 minutes. Proceed to iron-hematoxylin staining (step c)
    b. PVA/Modified PVA samples – mix the sample and pour onto a paper towel and let stand for 3 minutes (to absorb the PVA), take some of the sample and smear it onto 2 slides and allow to dry for several hours. Place the slides into iodine-alcohol for 1 minute (for PVA specimens to remove the mercury). Proceed to iron-hematoxylin staining (step c)
    c. Iron Hematoxylin Staining -Place the slides in 70% ethanol for 5 minutes
    d. Place slides in iodine-70% ethanol for 2-5 minutes
    e. Place the slides in 70% ethanol for 5 minutes
    f. Wash slides in running tap water
    g. Place the slides in iron hematoxylin working solution for 4-5 minutes
    h. Place the slides in running tap water
    i. Place the slides in 70% ethanol for 5 minutes
    j. Place slides in 2 changes of 100% ethanol for 5 minutes each
    k. Place slides in 2 changes of xylene for 5 minutes each
    l. Mount with coverslip using a mounting medium
    m. Allow the smear to dry and examine for protozoan throphs and cysts under oil

100x magnification
- Calcofluor white – for detection of Microsporidial spores and Acanthamoeba cysts
- Modified Acid fast stain – detects Cyclospora cayetanensis, Cryptosporidium, and Isospora spp.
- Weber Green Stain – for Microsporidia
- Ryan Blue stains – for Microsporidia
- Acid-fast trichrome stain – for Cryptosporidium and Microsporidia

4. Blood – stained with Wright's or Wright's-Giemsa stain

- Thin smear – prepared like that for a peripheral blood smear and provides an area of examination where the RBCs are neither overlapping nor distorted. The morphologies of parasites and infected RBCs are most typical
- Thick smear – utilizes more blood than the thin film. RBCs are lysed during staining making the preparation more or less transparent and leaving only parasites, platelets and WBCs for examination.

5. Cellulose Tape – most widely used for the detection of human pinworm. Adult female migrates out the anal opening and deposits the eggs on the perianal skin usually during the night. The eggs and occasionally the adult female worms stick to the sticky surface of the cellulose tape. The cellulose tape is submitted to the laboratory for direct examination of the pinworm and eggs

6. Other Specimens
- Sputum, bronchoscopy, transbronchial aspirate
- Skin, Skin scrapings, Biopsy
- Duodenal aspirate
- Urine (Schistosomas)
- Urogenital (vaginal discharge, urethral discharge, prostatic secretions)

SECTION IV
Parasite Identification

Ova and Parasite Examination
Stool ova and parasites exam is an analysis of mainly stool (other specimens less frequently) to check for the presence of a parasite or worm-like infection of the intestine. Ova refers to the egg stage of a parasite's life cycle. Some parasites are single-cell organisms such as amoeba, Giardia, and Trichomonas, while others have a worm-like appearance.

Scan the wetmount slides (0.85% saline slide first and then the Iodine slide) under 10x magnification and confirm any suspicious structures under 40x. In addition, at least 1/3 of the coverslip should be examined under the 40x magnification.

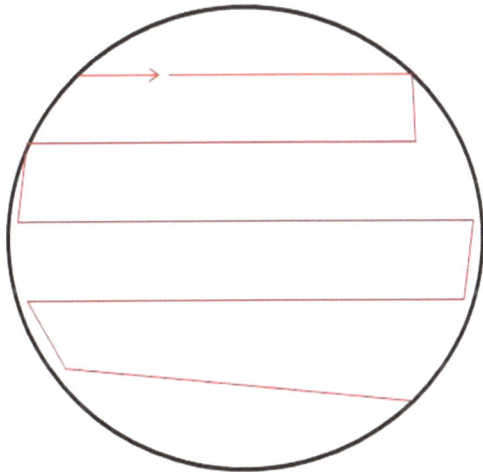

Picture 1 - Diagram on how to scan parasites in a microscopic field

Scan the slide in this manner to cover the entire slide

Key points to remember:

1.   Source
   - is very important, majority of parasites causing human disease is concentrated in the gastrointestinal tract, fecal matter is the most common specimen, check the list of common pathogens isolated in a specific source (Table 1) and this will narrow down the choices.
   - non-fecal specimen - source should be used to narrow down the list of organisms commonly isolated on those sources (refer to Table 1)

2.   Eggs
   - Some eggs are easily identified morphologically, always have an Atlas of the commonly isolated eggs handy or consult this manual for pictures of routinely encountered parasitic eggs

3.   Larvae
   - Check parasites that will normally produce larvae, always have an Atlas of the different larvae handy

4.   Worms
   - Organism's structures should be examined for reproductive system, head (scolex), presence of hooks, segments of the body (proglottid), tail, size
   Cellulose tape – check for presence of pinworm (Enterobius vermicularis) eggs or the actual worm

   Most clinical laboratory quantitates the result (on HPO/HPF) as either:

   - Rare/occasional, few, moderate or many
   - 1+, 2+, 3+ or 4+
   - 0-1/HPF, 1-3/HPF, 3-5/HPO, and so on

   Most clinical laboratory reports the result as quantitation + Name of parasite identified

Example: Rare Ascaris lumbricoides or
          1+ Ascaris lumbricoides or
          0-1/HPF Ascaris lumbricoides

Serological Tests
Useful for those parasites that invade deeper tissues. The following antibody detection tests are offered by the CDC

1.   Amoebiasis - Entamoeba hystolitica – EIA
2.   Babesiosis – Babesia microti, Babesia sp. – IFA
3.   Chagas disease – Trypanosoma cruzi – IFA
4.   Cysticercosis – larval Taenia solium – Immunoblot
5.   Echinococcosus – Echinococcus granulosus – EIA, immunoblot
6.   Leishmaniasis – Leishmania donovani, L. brasiliensis, L. tropica – IFA
7.   Malaria – Plasmodium vivax, P. ovale, P. malariae, P. falciparum – IFA
8.   Paragonimiasis – Paragonimus westermani – immunoblot
9.   Schistosomiasis – Schistosoma haematobium, S. mansoni, S. japonicum, Schistosoma spp. – Fast ELISA and immunoblot
10.   Strongyloidiasis – Strongyloides stercolaris – EIA
11.   Toxocariasis – Toxocara canis – EIA
12.   Toxoplasmosis – Toxoplasma gondii – IFA-IgG, EIA-IgM
13.   Trichinellosis (Trichinosis) – Trichinella spiralis – EIA

Blood smear - look for parasites inside the RBCs, inside the WBCs, or outside of the cells (extracellular)
INTRACELLULAR (INSIDE THE RBCs)

Plasmodium falciparum – infects young and mature RBCs – hemolysis is greatest

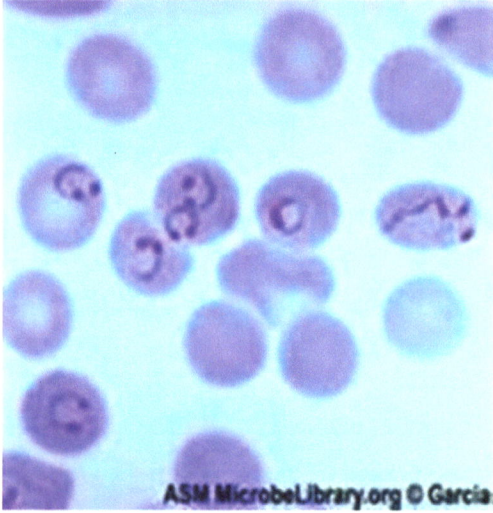

Picture 2 – Plasmodium falciparum multiple rings

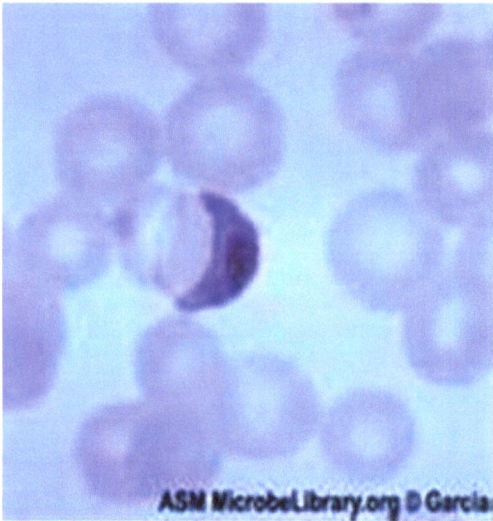

Picture 3 - Plasmodium falciparum gametocyte

On smear: RBCs normal in size. Ring form varies and may be missed because it is small, with small cytoplasmic circle and 1-2 chromatin dots. Trophs grow and is usually the main form seen in circulation. Schizont occurs in visceral organs and only rarely in peripheral blood. Crescent or banana-shaped gametocytes have deep blue cytoplasm with brownish pigment near center – easily recognized in thick smears

Plasmodium vivax and Plasmodium ovale – infect young RBCs and retics

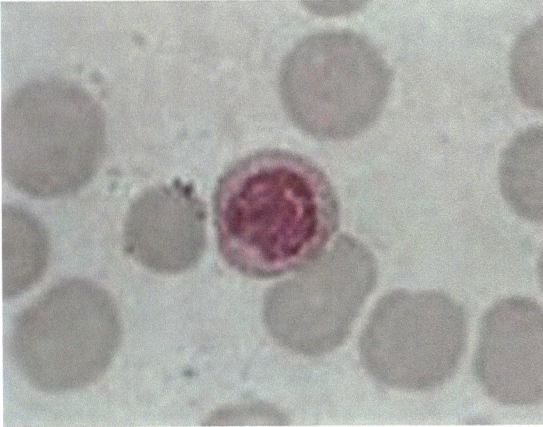

Picture 4 – Plasmodium vivax

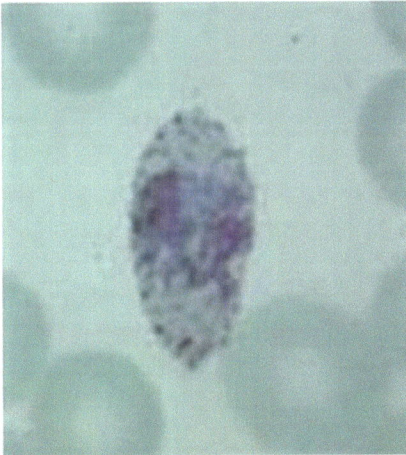

Picture 5 - Plasmodium ovale

Plasmodium Vivax – hemolysis is less
• On smear: trophozoites show a small ring with a red chromatin dot and a blue-staining cytoplasmic cycle→RBC grows larger, Schuffner stippling in late stages, ring shape is lost, the amoeboid throphs almost fill the cell→yellow-brown pigment from digested hemoglobins appear, division occurs (up 12-24 merozoites)→RBC burst

Plasmodium ovale –young trophs resemble P. vivax and P. malariae
• On smear: RBCs enlarges, oval, Schuffner stippling. On Schizont stage, RBCs are oval and parasite is round in the center of the cell. Merozoites can number 4-8. Gametocytes cannot differentiate from P. vivax, length of asexual cycle is 48 hours.

Plasmodium malariae – infects mature RBCs

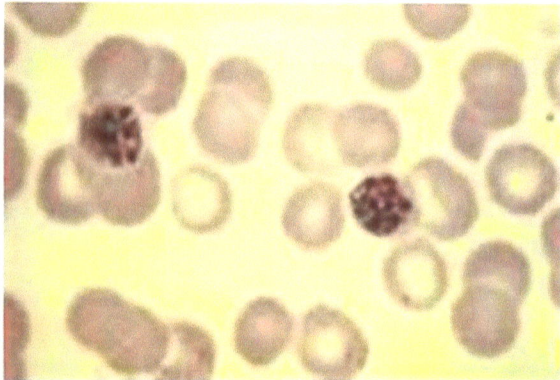

Picture 6 - Plasmodium malariae

• On smear: ring stage has a single chromatin dot and blue cytoplasmic ring that is often smaller and heavier than P. vivax but hard to distinguish. Growing trophs have chromatin that is round and streaky, cytoplasm is in a narrow band across the cell, course brown pigment is noted. RBCs do not become enlarged, no stippling, gametocytes similar to P. vivax but smaller, asexual cycle is 72 hours

Babesia microti

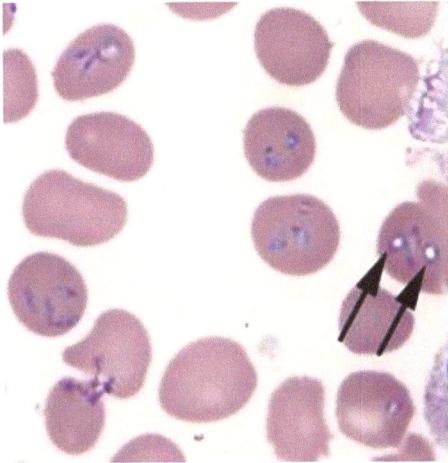

Picture 7 -Babesia microti

•   On smear: Small, ring-like protozoa within RBCs resembling stages of falciparum malaria. Ring-shaped parasites (thick smear preferred over thin due to scarcity of the parasite), tiny rings (1-4) with minute chromatin dots with minimal cytoplasm. Multiple rings (>2) differentiates it from falciparum

INTRACELLULAR (INSIDE TISSUES and WBCs)
Leishmania donovani
- On smear: oval amastigote form in tissues and inside WBCs during phagocytosis

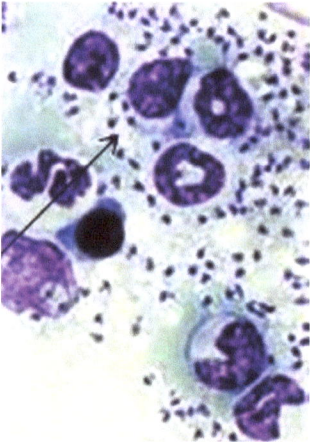

Picture 8 – L. donovani amastigote (in tissue)

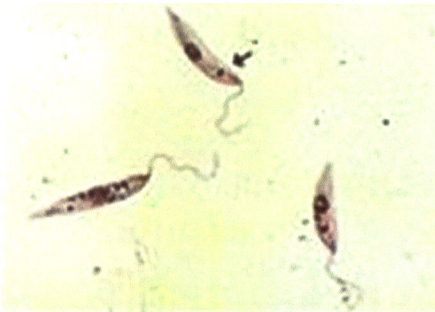

Picture 9 – L. donovani promastigote (in culture)

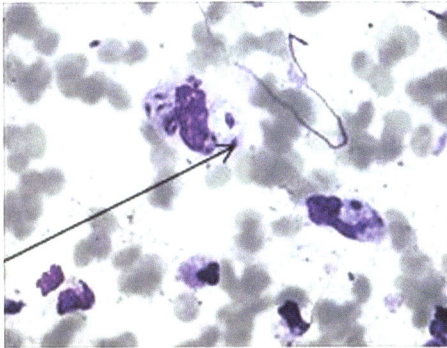

Picture 10 – L. donovani (in blood inside WBC)

Toxoplasma gondii
- On smear: intracellular in WBC during phagocytosis, in tissues will appear as crescent-shaped tachyzoites

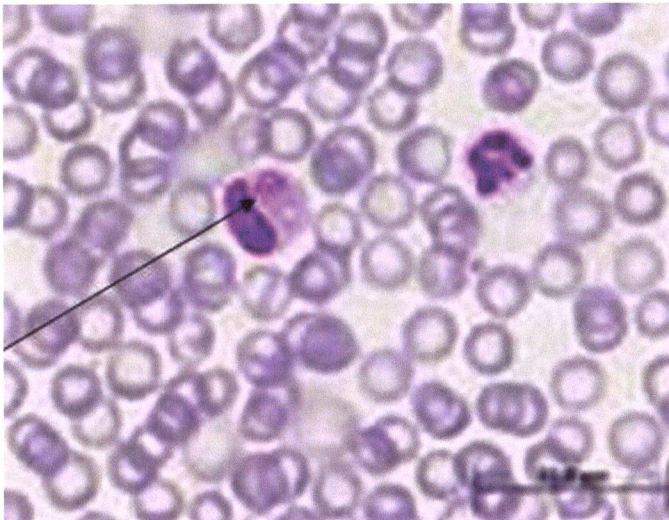

Picture 11 – Toxoplasma gondii in WBC (arrow)

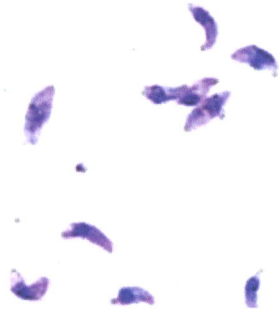

Picture 12 - T. gondii tachyzoites released from tissues

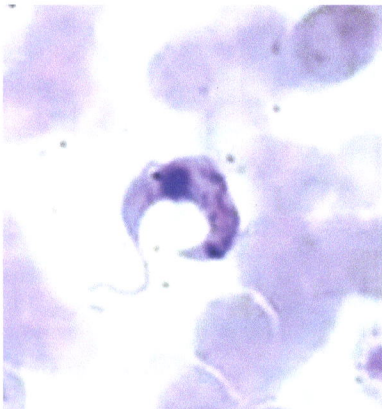

EXTRACELLULAR
Picture 13 - Trypanosoma cruzi trypomastigote

- On smear: C-shaped trypomastigote with terminal kinetoplast
Trypanosoma brucei gambiense and Trypanosoma rhodisiense

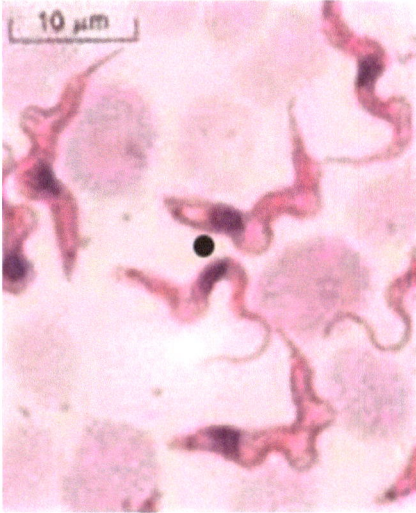

Picture 14 - Trypanosoma brucei gambiense

- On smear: Characteristic "C" or "U" shaped, narrow undulating membrane and large kinetoplast. T. brucei gambiense and T. rhodisiense looks morphologically the same on the smear

Microfilaria
- On smear: long thin nematode, identification is based on the presence or absence of the sheath and the distribution of the nuclei in the tail region

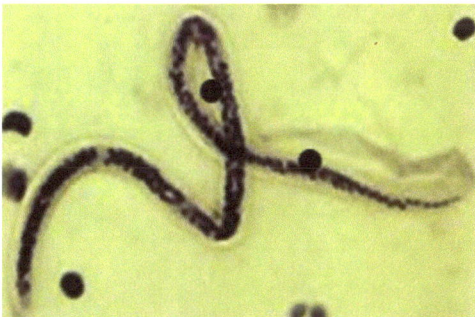

Picture 15 – Wucherera bancrofti

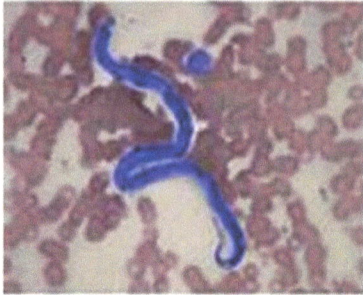

Picture 16 – Loa loa

Common eggs encountered in the stool specimen all under 40x except trichrome stains are under 100x:

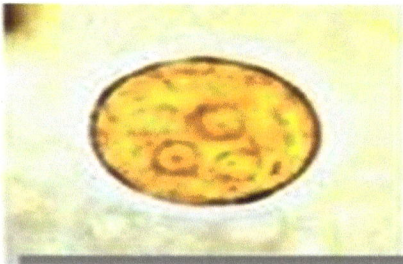

Picture 17 – Entamoeba hystolitica (cyst)

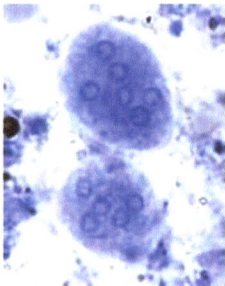

Picture 18 – Entamoeba coli (cyst)

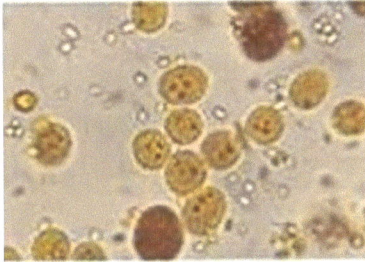

Picture 19 – Endolimax nana (cyst)

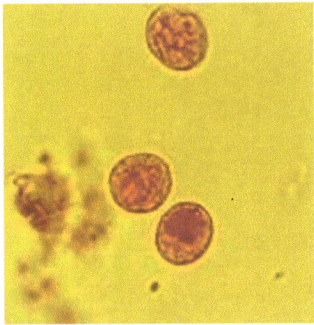

Picture 20 Iodamoeba butschlii (cysts)

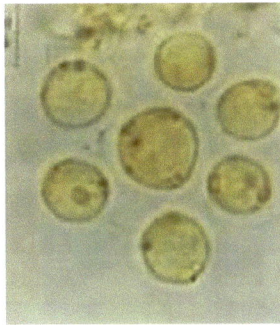

Picture 21Blastocystis hominis

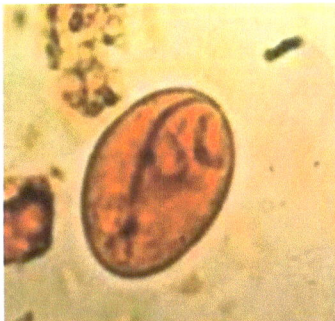

Picture 22 Giardia lamblia (cyst)
(cysts)

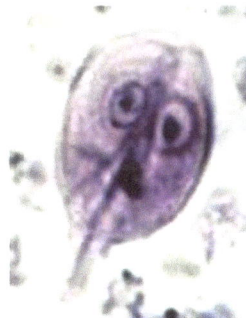

Picture 23 Giardia lamblia
(troph)-trichrome

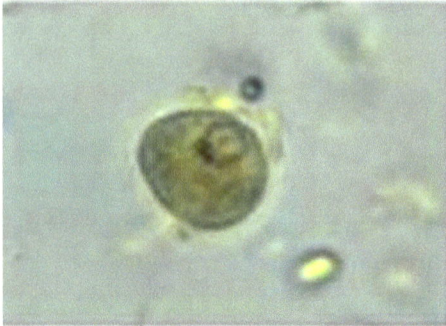

Picture 24 – Chilomastix mesnili (cyst)

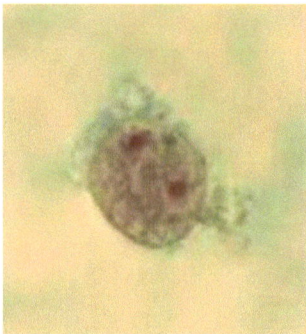

Picture 25 Dientamoeba fragilis   troph –trichrome

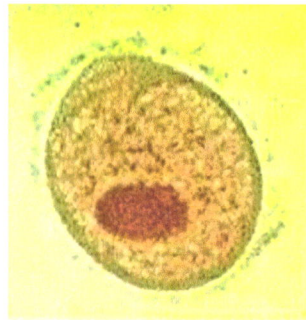

Picture 26Balantidium coli(troph)

Picture 27 – Isospora belli

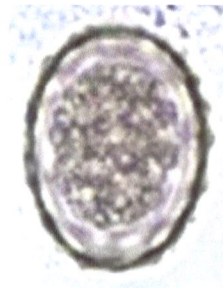

Picture 28 Ascaris lumbricoides (fertilized corticoid)

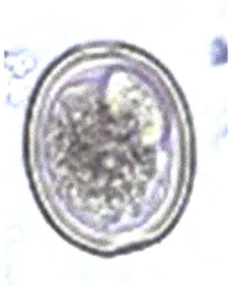

Picture 29 Ascaris lumbricoides (fertilized decorticoid)

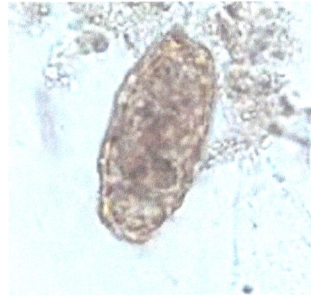

Picture 30 Ascaris lumbricoides (unfertilized corticoid)

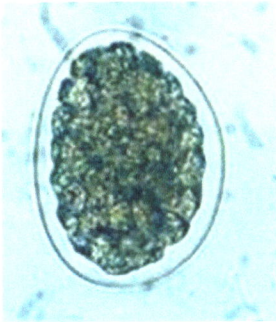

Picture 31 Hookworm eggs

(can't tell between the 2 hookworms)

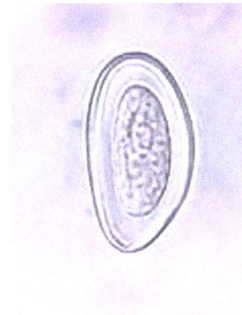

Picture 32 Enterobius vermicularis Pinworm

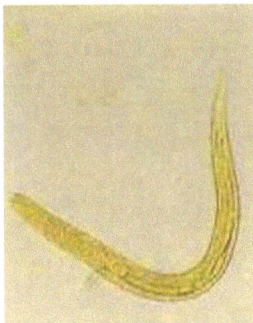

Picture 33 Strongyloides stercolaris (larva)

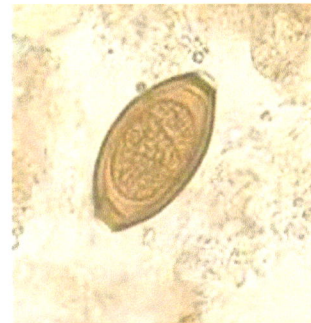

Picture 34 Trichuris trichiura -Whip worm

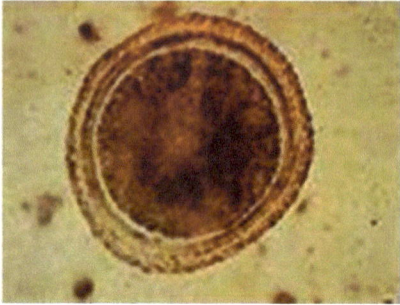

Picture 35 – Tapeworm

(hard to tell if beef or pork tape worm, examine scolex and proglottid)

Picture 36 - Diphyllobotrium latum (Fish tapeworm)

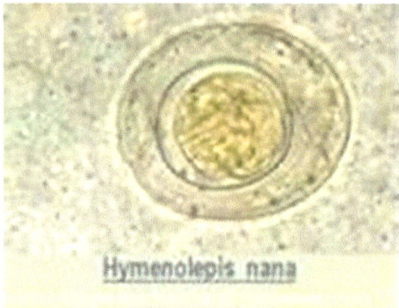

Picture 37 - Hymenolepis nana
(Dwarf tapeworm)

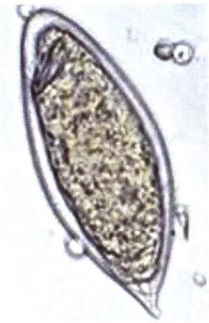
Picture 38 Schistosoma haematobium

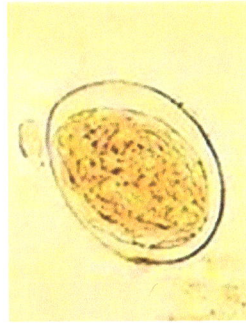
Picture 39 Schistosoma japonicum

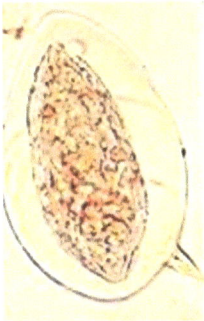
Picture 40 Schistosoma mansoni

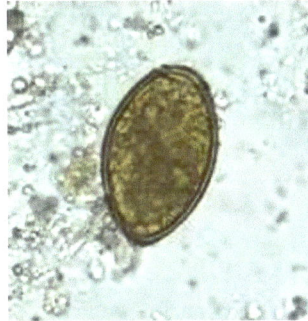
Picture 41 Paragonimus westermani

Picture 42 Clonorchis sinensis

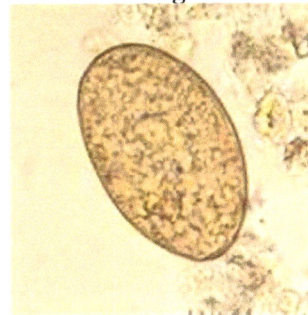
Picture 43 Fasciolopsis buski

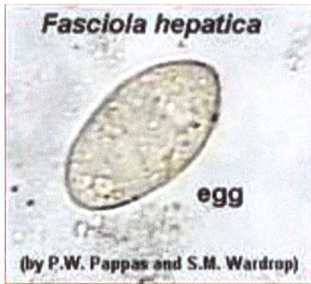

Picture 44 - Fasciola hepatica

LIVER/SPLEEN

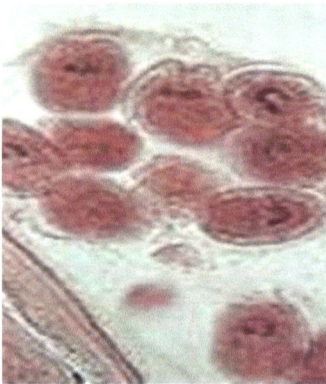

Picture 45 - Echinococcus granulosus

SECTION V
Medically Important Parasites

Plasmodium – causative agent of malaria, kills 1-3 million annually; mostly children. Transmitted by Anopheles mosquito

o Plasmodium vivax - worldwide, especially in the tropics, not as common in Africa, difficult to eradicate, relapse may occur, schizont may lie dormant for years in liver

o Plasmodium ovale – rare and found principally in Africa

o Plasmodium falciparum - most pathogenic human malaria, will run an acute fulminating course ending in death if treatment is not prompt. Massive intravascular hemolysis, hemoglobinemia, hemoglobinuria, jaundice, acute renal failure, Disseminated Intravascular Coagulation (gametocytes plug the circulation), cerebral malaria (due to sludging of damaged RBCs in microvasculature)

o Plasmodium malariae - Found in tropical Africa, India, Burma, Malaysia and Indonesia.8-12 merozoites. Treatment with chloroquine compared to P. vivax, rare nephropathy with albuminuria, hematuria and edema

- Clinical Syndrome of malaria:
  a. Fever, chills, rigors, sweating, headache, muscle pain, prostration, 25% with hemolytic anemia (jaundice, splenomegaly, hepatomegaly)
  b. Amount of hemolysis is related to number of infected RBCs
  c. Few may have DIC
  d. 75% of patients resistant to chloroquine resulting to thrombocytopenia
  e. Blood smear
     thick smear (water based Wrights-Giemsa) lyses RBCs to release the merozoites which can be observed microscopically
     thin smear – regular blood smear

Babesia

o Babesia microti – causes uncommon hemolytic disorder, transmitted from Ixodes damini and by blood transfusion
   - More severe in splenectomized individuals
   - Occurs in Nantucket island, on coastal regions of northeastern side of the U.S., in California, France, Ireland, Scotland, other European countries
   - Symptoms:
     a. 1-4 week incubation
     b. Onset of malaise, fever, chills, headache, drenching sweats, arthralgias, weakness and fatigue
   - Blood smear – look for organism inside RBCs
   - Serological tests - IFA

Trypanosoma

o Trypanosoma cruzi – agent of Chagas Disease or American Trypanosomiasis transmitted by the "Kissing bug" or bugs from the family Reduviidae (Reduviid bugs), found in areas extending from Central America to Southern Argentina, affects 16-18 million, the leading cause of heart disease and approximately 50,000 deaths annually
   - Reduviid bug is a large winged insect found in cracked walls and thatch of poorly constructed buildings, likes to bite near the eyes or lips of the sleeping host and defecate after feeding

- Appearance of nodular, erythematous chagoma (swollen eyelids), then progress to fever and other organ involvement
- Finding of organism inside WBCs or on tissues confirms diagnosis
- DNA tests
- Serological – IFA

○ Trypanosoma brucei gambiense - agent of African Sleeping Sickness (West and Central Africa) – the patient is in a comatose stage – transmitted by the Tse tse fly
○ Trypanosoma brucei rhodesience – agent of East African Sleeping Sickness transmitted by the Tse tse fly

Microfilaria
○ Wuchereria bancrofti – transmitted by Armigeres mosquito, causes elephantiasis
○ Loa loa – transmitted by a deer fly – causes Calabar swellings mainly of the eyes but also infects the skin
○ Brugia malayi – transmitted by mosquitos, milder form of elephantiasis
○ Onchocerca volvulus – transmitted by Simulium black flies, causes River Blindness, rash and other skin infection

Leishmania
○ Leishmania donovani – intracellular in the cells of the reticuloendothelial system (connective tissues) transmitted by the sandfly Phlebotomus spp. (Old World) and Lutzmyia spp. (new world) causes visceral leishmaniasis also known as

Kala Azar
○ Leishmania braziliensis – mucocutaneous leishmaniasis

Toxoplasma
○ Toxoplasma gondii – cat is the definitive host, causes mild flu-like illness but have a serious or fatal effect on the fetus, pregnant women with infected cats are at risk

Causes Gastrointestinal diseases
○ Entamoeba hystolytica – causes amoebic dysentery
○ Entamoeba coli – commensal parasite
○ Endolimax nana – considered non-pathogen
○ Iodamoeba butschlii – considered non-pathogen
○ Blastocystis hominis – associated with irritable bowel syndrome
○ Giardia lamblia – commonly isolated, causes diarrhea
○ Chilomastix mesnili – considered non-pathogen
○ Dientamoeba fragilis – infects large intestine, may be benign
○ Balantidium coli – found in cecum and colon, causes diarrhea
○ Isospora belli – found in tropical and subtropical areas, causes diarrhea
○ Ascaris lumbricoides – most common cause of parasitic diarrhea
○ Enterobius vermicularis – pin worm, causes diarrhea
○ Hookworm (Ancylostoma duodenale and Necator americanus) – infects > 600 million people worldwide – blood suckers causing anemia and diarrhea
○ Strongyloides stercularis –larva found in small intestine causing dermatitis, can migrate to the lungs
○ Trichuris trichuria – infects close to ¼ of the world's population, infects the cecum and the colon, may cause bloody diarrhea and causes vitamin A deficiency
○ Taenia

- - Taenia solium – pork tapeworm, causes cysticercosis a major neurological damage and may lead to death
    - Taenia saginata – beef tapeworm, infects cattle and humans, can become very large, up to 20 m, can develop into cystycercus, mainly cause diarrhea
  - Dyphillobotrium latum –fish tapeworm, causes gastroenteritis and anemia due to decrease or even severe Vitamin B12 deficiency
  - Hymenolepis nana – live in intestines of rats, ingestion of food contaminated by insects (beetles-secondary host) – causes abdominal pain, loss of appetite, diarrhea
  - Schistosoma – blood flukes, second in importance to malaria, transmission is through larval penetration of the skin
    - Schistosoma haematobium – urinary schistosomiasis, hematuria and fibrosis of the bladder
    - Schistosoma japonicum –endemic in Kayatam, Japan – if untreated may cause chronic hepatosplenic disease and impaired physical and cognitive development
    - Schistosoma mansoni – causes intestinal schistosomiasis, fever, cough and rash at the point of entry
  - Paragonimus westermani – lung fluke. Popular in Asia and South America-consumption of raw and undercooked seafood (crab species)
  - Clonorchis sinensis – Chinese liver fluke, found in the dwellings of the bile duct, consumes the bile and causes humans to ineffectively digest food, may cause biliary obstruction
  - Fasciolopsis buski – large intestinal fluke, infects amphibic snails and aquatic plants, ingested by pigs and humans, causes diarrhea, abdominal pain, intestinal obstruction
  - Fasciola hepatica –liver fluke or sheep liver fluke – causes liver damage
  - Cryptosporidium parvum – major cause of gaybowel syndrome, modified acidfast positive (acidfast-trichrome)
  - Cyclospora cayetanensis - contaminated raspberries, causes traveler's diarrhea

MUSCLE
- Trichinella spiralis – encystment in muscle tissues, ingestion of contaminated pork and other large game animals

UROGENITAL
- Trichomonas vaginalis – sexually transmitted, causes discharge and genital irritation

SECTION VI Treatment

ANTIPROTOZOAN AND ANTIHELMINTHIC DRUGS

- Antiprotozoan Drugs
  a. Quinine – used to control protozoan diseases such as malaria
  b. Chloroquine – synthetic derivatives, replaced quinine
  c. Mefloquine – recommended for chloroquine resistance but with serious psychiatric effects
  d. Quinacrine – recommended for giardiasis
  e. Diiodohydroxyquin (iodoquinol) – drug prescribed for several intestinal amoebic diseases, dosage must be controlled to avoid optic nerve damage
  f. Metronidazole (flagyl) – one of the most widely used antiprotozoan drugs, acts against parasitic protozoa and obligately anaerobic bacteria (Clostridium), drug of choice for Trichomonas vaginalis, used in the treatment of giardiasis and amoebic dysentery. Mode of action is to interfere with anaerobic metabolism

- Antihelminthic Drugs
  a. Niclosamide – 1st choice of treatment against tapeworm infections, inhibits ATP production under aerobic conditions
  b. Praziquantel – equally effective against tapeworm infections by killing the worms by altering the permeability of the plasma membranes, broad spectrum, recommended for treating several fluke-caused diseases, especially schistosoma, causes helminthes to undergo muscular spasms making them susceptible to the attack by the immune system
  c. Mebendazole and albendazole – broad spectrum, few side effects and drug of choice for treatment of many helminthic infections, mode of action is inhibit formation of microtubules in the cytoplasm which interferes with absorption of nutrients by the parasite
  d. Ivermectin – drug that is used primarily in livestock industry as an antihelminthic, results to paralysis of the worm, occasionally used for treatment of several human parasitic diseases and even treatment of scabies (skin mite) and head lice

Treatment Specific for Malaria:

    a. Chloroquine – destroys stages of malaria in RBCs except drug-resistant P. falciparum

    b. Quinine sulfate – for drug-resistant P. falciparum

    c. Pyrimethamine and sulfonamides – use in combination of Quinine sulfate in drug-resistant P. falciparum

    d. Primaquine – treat hepatic stages of vivax malaria, may cause G6PD deficiency

Treatment specific for Babesia:

    a. Clindamycin combine with quinine shows success within 1 week

    b. Exchange transfusion – to treat transfusion-transmitted Babesia infection

    c. Chloroquine – treatment failure may occur

PARASITOLOGY -WRITTEN EXAMINATION
(20 POINTS)

Student Name_____ Date_____
(Please circle the correct answer)

1.   The following are flukes except:

    a.   Paragonimus westermani

    b.   Fasciola hepatica

    c.   Trichuris trichura

    d.   Schistosoma japonicum

2.   Wetmount examination of this will assess the motility of the parasite:

    a.   Lugol's iodine

    b.   Saline

    c.   D'Antoni's Iodine

    d.   Formalin-ethyl acetate

3.   The following are roundworms except:

    a.   Echinococcus granulosus

    b.   Ascaris lumbricoides

    c.   Trichinella spiralis

    d.   Enterobius vermicularis

4. This amoeba is the causative agent of amoebic dysentery and transmitted between humans through ingestion of cysts:

    a. Entamoeba coli

    b. Giardia lamblia

    c. Entamoeba dispar

    d. Entamoeba hystolitica

5. Animals characterized by segmented bodies, hard external skeletons and jointed legs and belongs to the largest phylum in the animal kingdom where a few can suck the blood of humans and animals and can transmit microbial disease:

    a. Rhizopods

    b. Arthropods

    c. Protozoans

    d. Helminthes

6. Match the vector with the parasite (1 point each) :

_____Babesia microtti   a. Phlebotomus spp.

_____Loa loa   b. Tse tse fly

_____Trypanosoma spp.   c. Ixodes damini

_____Plasmodium spp.   d. Riduvid bug

_____Leishmania donovani   e. Anopheles mosquito

_____Trypanosoma cruzi   f. Deer fly

7.   The following are O & P concentration methods:

   a. Sedimentation

   b. Centrifugation

   c. Flocculation

   d. Flotation

   e. a and b

   f. a and c

   g. a and d

   h. b and c

   i. b and d

   j. c and d

8.   The following parasites will stain positive with modified acid fast stains:

   a. Toxoplasma gondii

   b. Cyclospora cayetanensis

   c. Cryptosporidium parvum

   d. Trypanosoma brucei ganbiense

   e. a and b

   f. a and c

   g. a and d

   h. b and c

   i. b and d

   j. c and d

9.   The following are tapeworms except:

   a.   Taenia saginata

   b.   Diphyllobotrium latum

   c.   Hymenelopis nana

   d.   Opistrochis (Clonorchis) sinensis

10.   This parasite does not have a cyst stage so the transfer is from host-to-host and transmitted sexually or from contaminated toilet facility or towels:

   a.   Endolimax nana

   b.   Balantidium coli

   c.   Trichomonas vaginalis

   d.   Pentatrichomonas hominis

11.   The most pathogenic human malaria that will run an acute fulminating course causing massive intravascular hemolysis, DIC and will end in death if left untreated

   a.   Plasmodium ovale

   b.   Plasmodium falciparum

   c.   Plasmodium vivax

   d.   Plasmodium malariae

12.	This microfilaria is transmitted by the Armigeres mosquito causing the most severe form of elephantiasis:

　　a.	Loa loa

　　b.	Brugia malayi

　　c.	Wucherera bancrofti

　　d.	Onchocerva volvulus

13.	This pathogen is commonly isolated from the muscle tissues:

　　a.	Taenia saginata

　　b.	Trichomonas spiralis

　　c.	Taenia solium

　　d.	Trypanosoma cruzi

14.	The only ciliate that is a human parasite, causing severe but rare type of dysentery and transmitted by the ingestion of the cyst

　　a.	Balantidium coli

　　b.	Toxoplasma gondii

　　c.	Paragonimus westermani

　　d.	Trichinella spiralis

15.   This parasite is intracellular in the cells of the reticuloendothelial system, transmitted by a sandfly causing visceral infection known as Kala Azar disease

    a.   Leishmania tropicalis

    b.   Leishmania braziliensis

    c.   Leishmania donovani

    d.   Leishmania Mexicana

16.   Which O & P preservative contains mercuric chloride?

    a.   Modified PVA

    b.   SAF

    c.   MIF

    d.   PVA

17.   This amoeba causes Primary Amoebic Meningoencephalitis or PAM

    a.   Acanthamoeba spp.

    b.   Naegleria spp.

    c.   Echinococcus granulosus

    d.   Opisthorchis sinensis

18.   This fish tapeworm causes gastroenteritis and anemia with severe vitamin B12 deficiency in humans:

    a. Diphylobotrium latum

    b. Taenia solium

    c. Hymenelopis diminuta

    d. Taenia saginata

19.   This blood fluke causes urinary infection causing hematuria and fibrosis of the bladder

    a. Schistosoma mansoni

    b. Schistosoma haematobium

    c. Schistosoma japonicum

    d. Schistosoma indicum

20.   This anti-protozoan drug is one of the most widely used and acts against parasitic protozoa and the drug of choice for Trichomonas vaginalis, giardiasis, and amoebic dysentery

    a. Quinine

    b. Praziquantel

    c. Flagyl

    d. Mebendazole

Student Signature_____   Date_____

Total Correct Score_____

%_____

Answer Key Parasitology Written Examination
20 Points

1.	c
2.	b
3.	a
4.	a
5.	b
6.	c,f,b,e,a,d
7.	i
8.	h
9.	d
10.	c
11.	b
12.	c
13.	b
14.	a
15.	c
16.	d
17.	b
18.	a
19.	b
20.	c

Afterword

In a perfect world, after their clinical rotation, CLSs and MLTs will be working in all four areas of study of medical technology (Microbiology, Chemistry, Hematology and Immunohematology), retain all that they know and live and work happily ever after. But as we all know that is far from the case, more frequently than not, a CLS or MLT will be stuck in one or two specialized area of study. Parasitology is a unique sub-subspecialty of Microbiology that the microscopic part of it requires years of experience for a CLS to feel a high level of confidence. CLSs who had been away from the Parasitology department for a period of time or never had a chance to work in the Parasitology department will need a good refresher course before venturing into this department. This manual does not claim to be able to boost someone's confidence overnight or claim to have all the answers, but instead this manual serves as a guide to re-discovering what one previously knew.

My hope is that this manual will serve its purpose and be a source of confidence to those who are brave enough to venture out in the parasitology department as a newbee or someone who had been away from it for a period of time.

Glossary

This manual is directed towards CLS and MLT students at or nearing their clinical rotation as well as CLSs that had been away from the area of Parasitology for a period of time and should already be familiar with most of the terms used in this manual. Only a selected number of terms below were chosen for further explanation.

AMASTIGOTE – a phase in the life cycle of trypanosome protozoans when the cell does not have any cilia or flagella.

ARTHRALGIA – pain in one or more joints.

BRONCHOSCOPY – a medical procedure wherein a flexible device, an endoscope is used to inspect the bronchi by obtaining tissue specimens for examination.

CDC – Center for Disease Control and Prevention, a major operating component of the U.S. government's Department of Health and Human Services concerned with disease prevention and occupational health and safety.

COMMENSAL – an organism deriving food and benefits from another organism without hurting or helping the other organism.

DIC – refers to Disseminated Intravascular Coagulation, a pathological activation of coagulation mechanisms that happens in response to a variety of diseases. DIC leads to the formation of small blood clots inside the blood vessels throughout the body.

DORMANT – inactive stage of an organism.

EIA – Enzyme Immunoassay, is a laboratory technique to detect the presence of an antigen in a sample by employing a series of biochemical steps.

ELISA – Enzyme-linked Immunosorbent Assay, is a laboratory technique to detect the presence or an antigen or antibody in a sample by employing a series of biochemical steps.

FULMINATING – rapid, sudden and severe such as an infection, fever or hemorrhage.

GAMETOCYTES – refers to the protozoan cell that divides through mitosis to produce gametes.

HEMOGLOBINEMIA – refers to a condition where there is an excess free hemoglobin in the plasma.

HEMOGLOBINURIA – refers to a condition where there is an excess of free hemoglobin in the urine.

HERMAPHRODITIC – a condition where the organism possesses both male and female reproductive organs.

IFA – Indirect Fluorescent Antibody, is a laboratory technique to detect the presence of antibodies to a specific antigenic material using fluorescent tagged antibodies to visualize the reaction under a fluorescent microscope.

IMMUNOBLOT – a laboratory procedure in which proteins that had been separated by electrophoresis are transferred "blotted" onto a nitrocellulose sheets and identified by their reactions with labeled antibodies.

KINETOPLAST – is a round mass of extranuclear DNA-containing organelle of kinetoplastid protozoans that is usually found in an elongated mitochondrion located adjacent to the basal body.

MEROZOITES – a small amoeboid sporozoan trophozoite produced by schizogony that is capable of initiating a new sexual or asexual cycle of development.

MITOCHONDRIA – any of various round or long cellular organelles of most eukaryotes that are found outside of the nucleus and function as energy producers.

MUCOCUTANEOUS – skin and mucous membrane involvement.

SCHIZONT STAGE – is a stage in the malarial life cycle where the red blood cells contains round inclusions that contain deeply staining merozoites.

SCHUFFNER STIPPLING – also known as Schuffner's dots or Schuffner's stippling, are stainable granules on malarial infected red blood cells.

SYMBIONTS – refers to the smaller or the beneficiary in organisms in a relationship of living together without harming each other.

TRYPOMASTIGOTE – any flagellate of the family Trypomastidae that has the typical form of a mature blood trypanosome.

UROGENITAL – refers to the organs of excretion and reproduction.

VISCERAL – internal organs of the body especially referring to those organs in the large cavity of the trunk such as the heart, liver and intestine.

References

Clinical Parasitology. 9th Edition. Beaver, PC, Jung, RC, Cupp, EW. Philadelphia. Lea and Febiger, 1984.

Diagnostic Medical Parasitology. 5th Edition. Garcia, LS. Washington, D.C.. ASM Press, 2007.

Medical Chemical Corporation Ova and Parasite Transport Kits Manufacturer Inserts.

Merriam-Webster, m-w.com

About the Author

Mary Michelle Shodja earned her Bachelors of Science Degree in Medical Technology in 1992 from California State University (CSU) Dominguez Hills in Carson, California. She took her medical technology clinical year training from the Southern California Kaiser Permanente Medical Hospital and Regional Laboratory. She earned her Masters of Science Degree in Bioanalysis in 1995 from her alma mater CSU Dominguez Hills.

She gained Certifications in both the MLS American Society of Clinical Pathologists (ASCP) and CLS National Credentialing Agency (NCA) in 1993. Immediately after passing her California License, she started as a Staff CLS in the Microbiology Department of the Southern California Kaiser Permanente Regional Laboratory. For over 17 years as a CLS, she maintained a full-time position in the area of Microbiology and a part-time position as a generalist working in Hematology, Chemistry, Serology, Immunology and Non-transfusion Blood Bank.

Over the years, she took on managerial, supervisory, teaching and other administrative and consultative work but her real passion lies in the clinical bench work and teaching. She believes that in order to find what you're looking for, you have to try out everything else. She is currently awaiting admission to the PhD program in a number of California institutions desiring to earn her last educational degree, a PhD in Molecular Microbiology. She believes that the answers to the future research lies in the field of Molecular Biology and Genetic Engineering. She hopes to be involved in the various HIV research in the future.

www.ingramcontent.com/pod-product-compliance
Lightning Source LLC
Chambersburg PA
CBHW041450210326
41599CB00004B/200